JUNIOR FUNFAX ™

First Facts

OUR
WORLD

Written by Brian Jones
Illustrated by Brian Hoskins

HENDERSON
PUBLISHING PLC

©1993 HENDERSON PUBLISHING PLC

How Our World Was Made

Where do you live?

We call this world where we all live **Earth**. Earth is a planet,
a round ball of rock and liquid which hangs in space.
There are eight more planets close to Earth, and one star.
You already know the name of that star - we call it the
sun. Earth is the third planet away from the sun.
These nine planets make our **solar system** and
each one spins around the sun in a
huge circle called an **orbit**.

The sun is a star in our **galaxy**.
A galaxy is a huge swirl of
stars, millions of them,
floating in space.
There are millions of
galaxies dotted
throughout space.
We call our galaxy
The Milky Way.

The Solar System

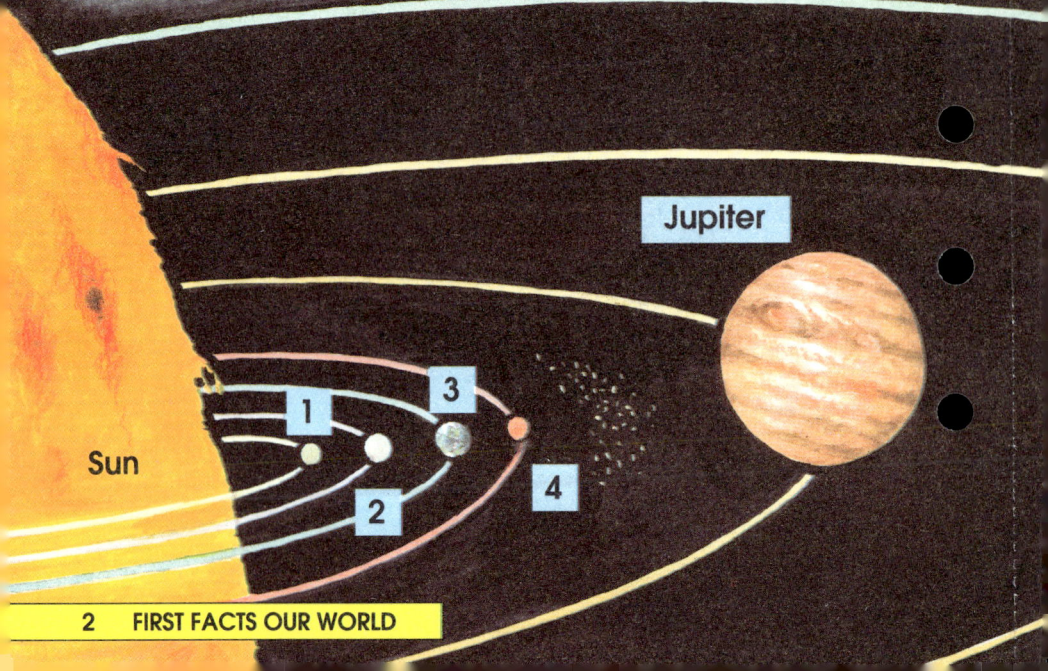

Jupiter

Sun

1

2

3

4

The solar system

Your life probably began about eight years ago. The sun and planets began theirs about 4,600 million years ago!

They all came from a huge cloud of dust and gas. Most of the gas at the centre of the cloud made the sun. The dust collected with the rest of the gas to make the planets.

How to build a planet

The four planets closest to the sun are called **Mercury** (1), **Venus** (2), **Earth** (3) and **Mars** (4). They are all rocky and were made up from the dust. In the beginning, they were very hot. As they cooled down, thin rocky crusts formed on the outside of them.

The other planets, **Jupiter, Saturn, Uranus** and **Neptune**, are all much bigger than Earth. They are called gas giants and have rocky centres. They are surrounded by deep layers of gas.

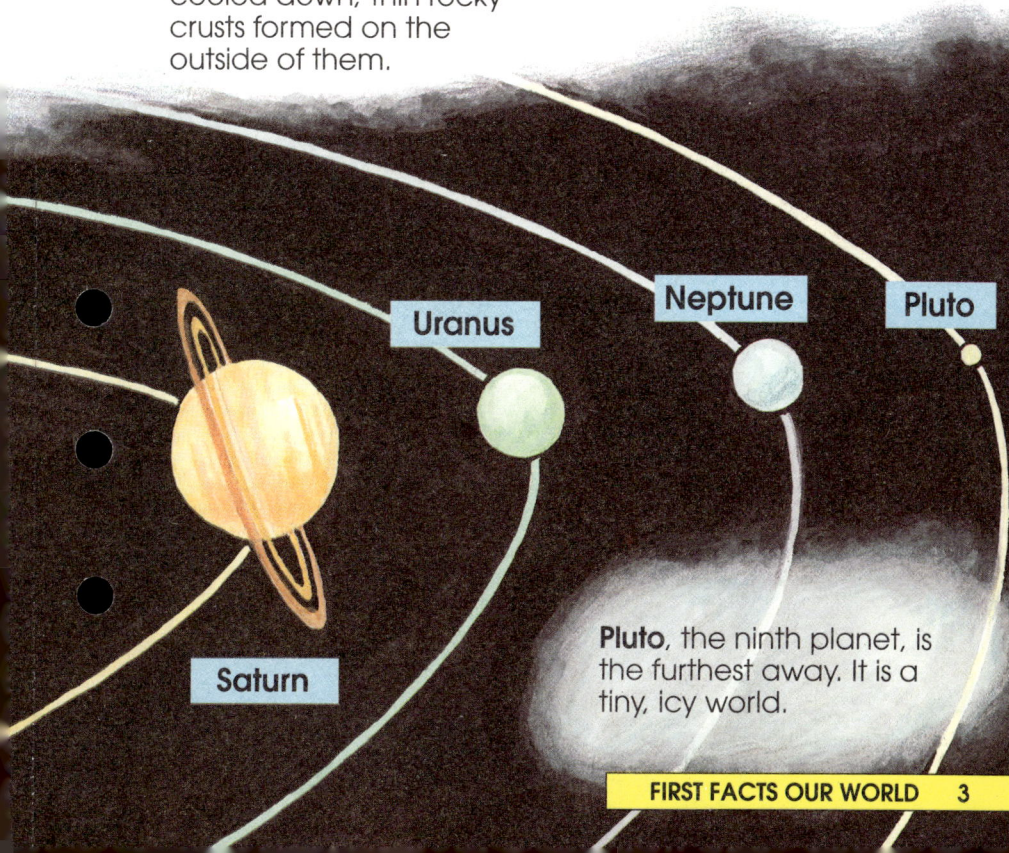

Uranus

Neptune

Pluto

Saturn

Pluto, the ninth planet, is the furthest away. It is a tiny, icy world.

Inside Our World

The Earth's crust

The Earth isn't quite round. It's a huge ball, flattened at the top and bottom, at the North and South Poles. It bulges out at the middle, which we call the Equator.

When astronauts look down on Earth from space, they see the **crust**, or outside. This thin layer of solid rock is the Earth's outer skin.

The Earth's crust is not the same thickness all over. Under the oceans, it is much thinner than on land.

This layer of the Earth is where you'll find coal, slate, gas and oil. We mine for coal and slate and drill for gas and oil.

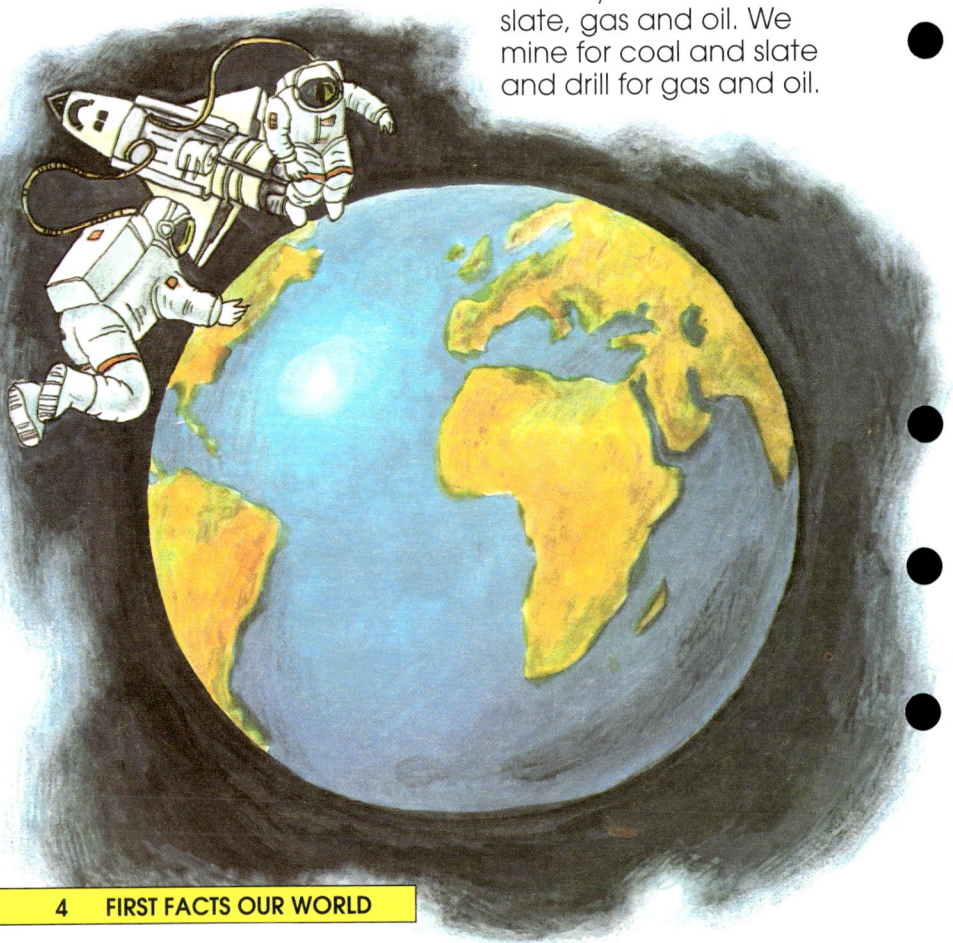

And deeper inside ......

Below the Earth's crust is the **mantle**. This is a very hot layer of hot liquid rock which flows about like hot, runny tar. The top layer is called the **magma**.

The Earth's crust has slowly broken up into pieces. We call these **plates**, floating about on top of the crust. The land lies on the plates. As they move around, the surface of our planet is constantly changing.

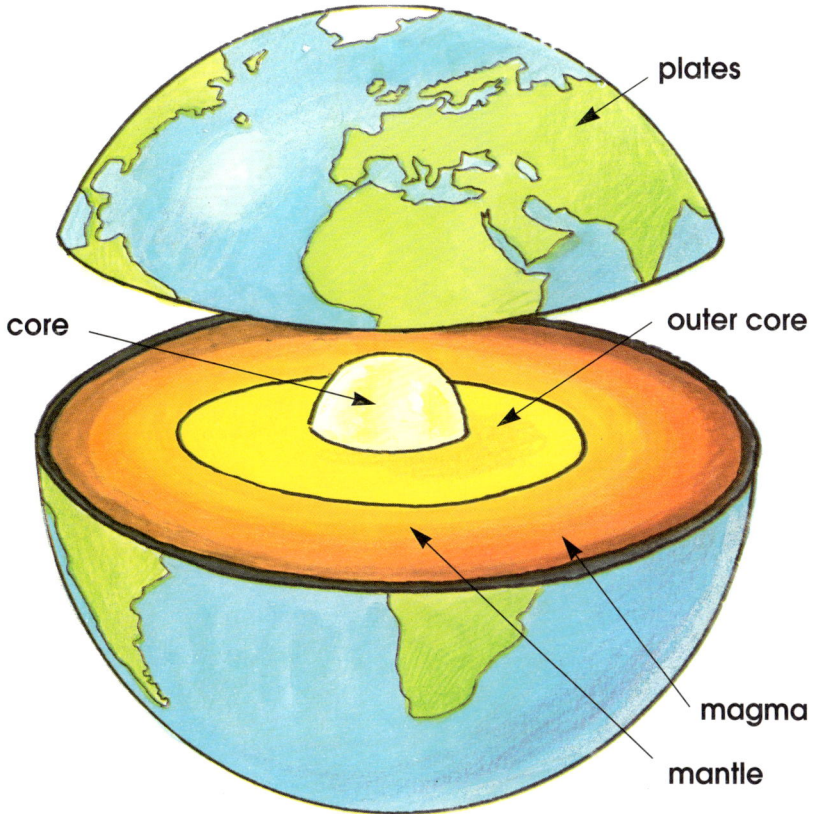

plates

core

outer core

magma

mantle

And deeper still ....

Deep inside the Earth there is a metal **core**. It is very, very hot - around 3,700 degrees Celsius.

Around the core is a liquid metal **outer core**. This contains minerals like iron and nickel.

A Changing Earth

Continents on the move

Look at a world map. Can you see seven big pieces of land? These are called **continents**. Which one do you live on?

Earth did not always look like this. Once, long ago, all the land was joined in one large continent.

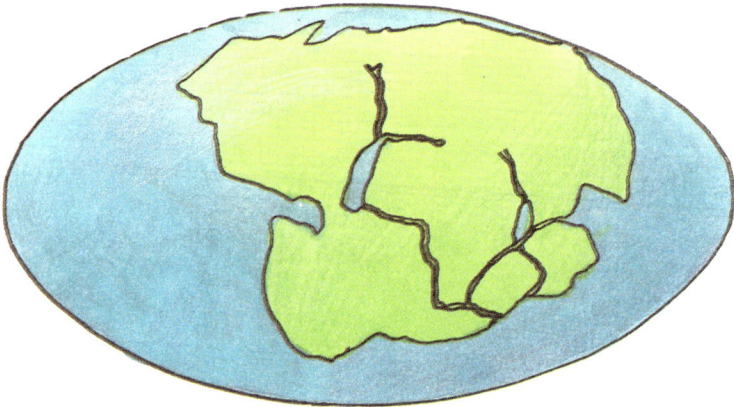

As the Earth's crust began to move, the land broke into two pieces. Then more pieces broke away. Slowly these pieces floated across the Earth's surface to make the continents as they are today.

Still on the move

Deep under the Atlantic Ocean, hot liquid rock is being pushed up through a deep crack in the Earth's crust. The rock forms a new crust and forces the two **plates** of old crust apart. The plates carrying North America and Africa move a few centimetres away from each other every year.

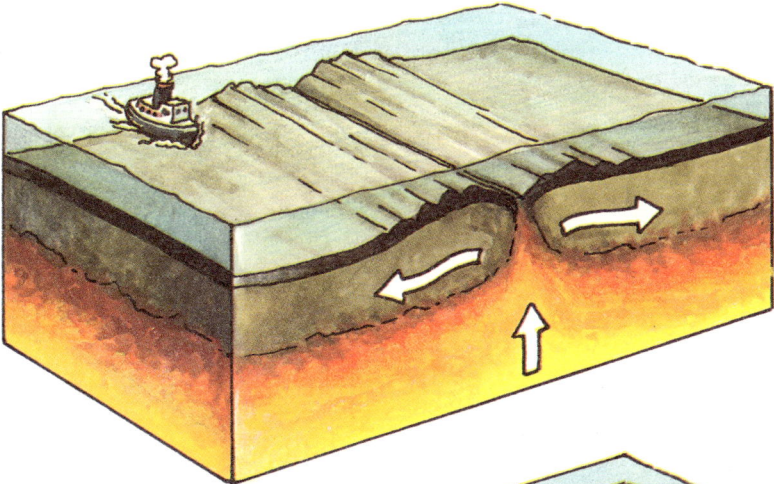

Ocean trenches

Sometimes plates can slide under each other. The old crust is pushed underneath. This is how deep ocean trenches are formed. Most of these trenches are found under the Pacific Ocean. These are the deepest parts of the ocean floor.

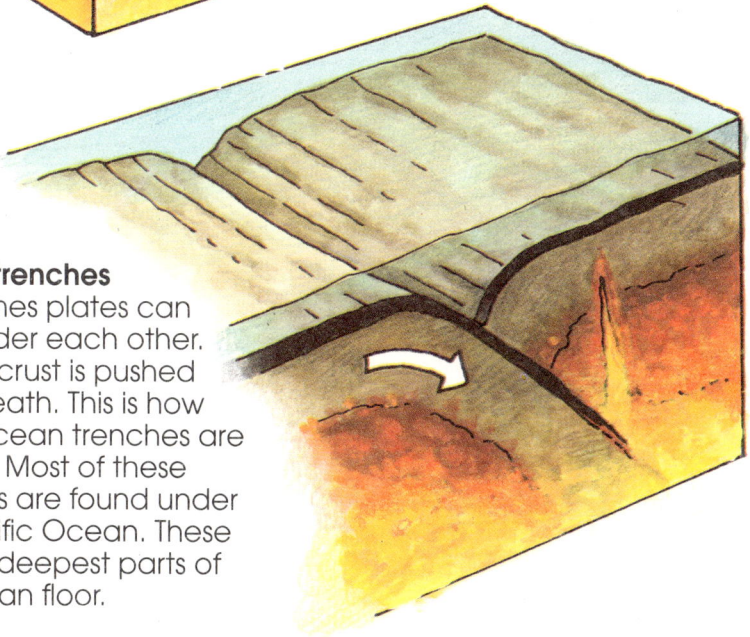

Mountains

Mountain chains

Plates can also squeeze together. They may fold over each other. This pushes them up and a mountain chain is formed.

There are many mountain chains on the Earth's surface. The longest mountain range is the Andes in South America.

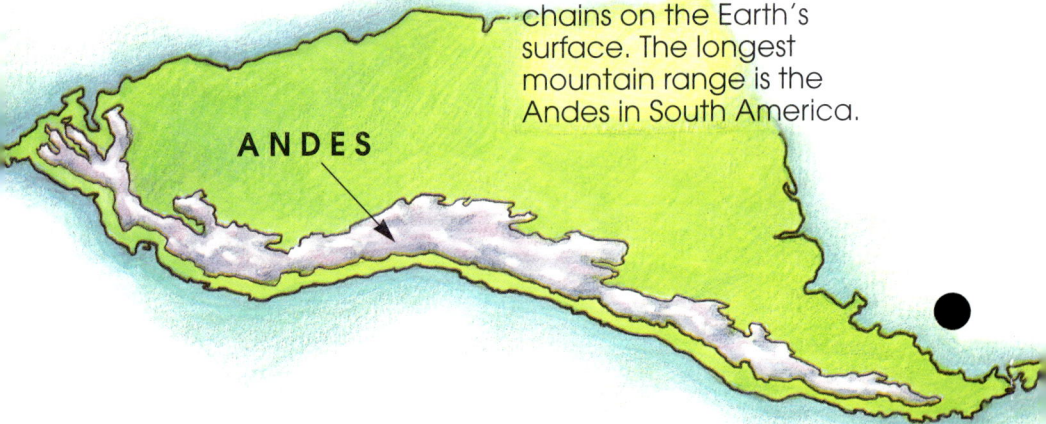

ANDES

The tallest mountain range are the Himalayas in Asia. 50 million years ago, the piece of the Earth's crust carrying India pushed into the plate carrying Asia, and these mountains were formed.

Making mountains

Another way of making mountains was to ooze hot liquid rock through the Earth's crust. The rock cooled and hardened and formed a mountain.

Record mountains

The highest mountain of all is Mount Everest. This is in the Himalayas. Mount Everest is nearly nine kilometres high.

There is a bigger mountain in Hawaii. This is called Mauna Loa. Mauna Loa is nearly ten kilometres high but nearly half of it is under the sea.

Volcanoes

What is a volcano?

A volcano is a hole in the Earth's crust. Hot liquid rock and lava from inside the Earth burst through the hole. Ashes and hot gases also gush out.

Some volcanoes explode violently, shooting rocks and lava high into the air. Mount Vesuvius in Italy is like this.

Others rumble away quietly. Lava pours out slowly over the rim. The lava is very hot. It flows down the sides of the volcano, burning up everything in its path.

Where are the volcanoes?

There are nearly 600 active volcanoes on Earth. These are both on land and under the sea. Many volcanoes are found around the edges of the Pacific Ocean.

Buried city

In 79AD, the ancient city of Pompeii, in Italy, was completely buried under hot ashes and stones when nearby Mount Vesuvius erupted.

Krakatoa cracks

In 1883, the volcanic island of Krakatoa, in the Pacific Ocean, erupted. Many, many people died. On nearby islands huge waves caused by the explosion killed many more people.

Record volcanoes

The highest volcano on Earth is in Chile. It is called Guallatiri and is over six kilometres high. But the biggest volcano of all is actually on Mars. Its name is Olympus Mons. It measures over 500 kilometres across its base and is 25 kilometres high!

The Atmosphere

The air you breathe

All around our planet, there is air. We call this **atmosphere**. It is made of different gases.

Oxygen is a very important gas because all animals, like us, need it to breathe.

Even though there are lots of people and animals on the Earth, we won't run out of oxygen. Our bodies can change the oxygen we breathe in into another gas called **carbon dioxide**. The carbon dioxide we breathe out is taken in by plants. They give it out again as oxygen! So oxygen is always being replaced.

Keeping us safe

The atmosphere also helps to keep out harmful rays from the sun. If it wasn't for the air protecting us, nothing could live on our planet.

It helps to keep out the sun's fierce heat during the day. At night, the air acts as a giant blanket and keeps in the heat we get from the sun.

The air is thickest at the Earth's surface, where we live. The higher we go, the thinner the air becomes. At the top of high mountains the air is very thin. Mountaineers need special oxygen supplies to allow them to breathe.

Airspace

There are many things in the air. There is also dust. Most of this comes from volcanoes which erupt and throw dust and ash high into the atmosphere.

Of course, we also see man-made objects in the air, including balloons and planes!

The Weather

Every cloud has...
In the atmosphere, groups of thousands of tiny specks of water float. These are clouds. There are many different kinds of clouds for you to name.

Cirrus clouds are very high. These are wispy and are made up of ice particles.

Stratus clouds are the lowest. These look like large sheets covering the sky.

Fog and mist are not called clouds because they are on the ground.

Rain and snow
The sun's heat makes water rise from the land and sea as a mist or fog. The water goes high into the air and collects into clouds. As the small specks of water join together they become heavier. This makes them fall as **rain**. The rain drains off the land into streams and rivers. At last, it weaves its way back to the sea.

All clouds are made up of water, but only a few make rain. If a cloud goes into warmer air, it **evaporates**, or fades away. Clouds in very cold air turn into ice particles and fall as snow.

Windy weather

The **Equator** is a line around the middle of the Earth. There isn't really a line to see, it is an imaginary circle. The sun heats air near the Equator and, because hot air rises, this causes winds.

This warmed air rises and moves away from the Equator. It cools down and then sinks again. After sinking, it either returns to the Equator or moves to the North and South Poles. All of this air flowing from one place to another produces wind. The faster the air moves, the stronger the winds it makes. Some winds are very strong. These are called **gales** and can cause a lot of damage.

The Seasons

Change of seasons

You may be surprised to know that our planet doesn't sit upright. The Earth is tilted at an angle. As it goes around the sun, one half is tilted towards it. In June and July, the northern half leans towards the sun. It is then summer in the north, in Europe for example. At the same time, the southern half is tilted away from the sun. It is winter in the south, like in Australia.

The seasons slowly change as the Earth moves around the sun. Six months later, the Earth is on the other side of the sun. Now, the northern half is tilted away from the sun. It is then winter in the north and summer in the south. There are no seasons in places near the Equator. This is because they are always facing the sun.

Oceans

A watery world

Three-quarters of the Earth's surface is covered by water. Most of this is in the Pacific, Atlantic, Indian, Arctic and Southern Oceans. The oceans are all linked together and form one huge expanse of water surrounding the Earth.

The biggest ocean is the Pacific. This huge area of water is larger than all of the Earth's land masses put together.

Undersea valleys are called **trenches**. The deepest trench of all is called the Mariana Trench. This lies under the Pacific Ocean and is over 11 kilometres deep!

Deep water

The ocean floors have mountains and valleys, just like on land. Some of the mountains on the ocean floors are so huge that they stick out above the water. These are our **islands**.

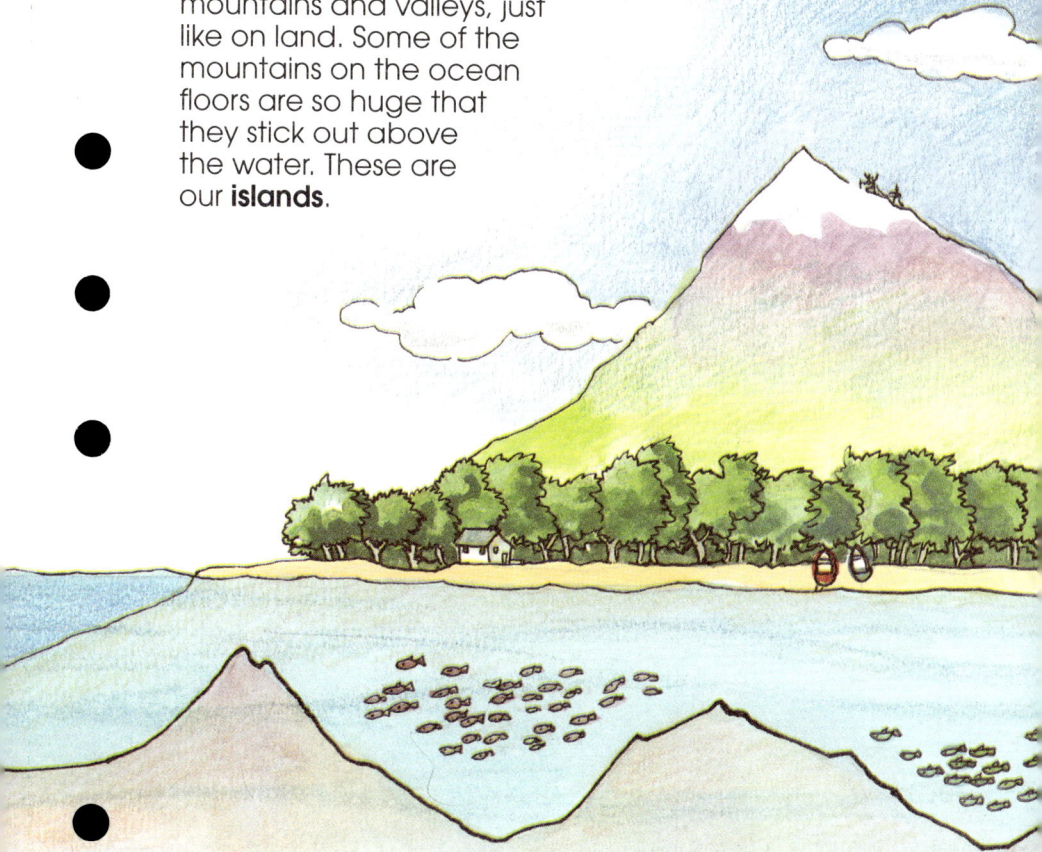

Don't drink the water

The water in the seas and oceans has salt in it. Some seas are saltier than others. The saltiest sea of all is the Dead Sea.

There is enough salt in the world's oceans to cover the whole of the Earth's land areas with a layer over 100 metres thick!

The Ice Caps

The Arctic and the Antarctic

The **Arctic** is a huge block of ice surrounding the Earth's North Pole. This ice floats on the Arctic Ocean.

The area around the Earth's South Pole is called **Antarctica**. Unlike the Arctic, Antarctica is not made up completely of ice. It is a mountainous continent covered by a deep layer of ice. Antarctica is much larger than the Arctic.

The areas around the poles are very cold.
The temperature in Antarctica can drop to 70 degrees Celsius below freezing point! The few animals that live in the polar regions have special thick coats or skins. Polar bears, seals and walruses live near the Arctic. Seals, penguins, some seabirds and whales live in Antarctica.

Icebergs

Icebergs are chunks of floating ice, some of them huge. During the summer, the ice blocks of the Arctic and the Antarctic warm up and chunks break off.

Arctic icebergs float into the Atlantic Ocean. They can be a danger to ships and boats. In 1912, a huge passenger liner, Titanic, sank after striking an iceberg near Newfoundland.

Record Icebergs

The largest iceberg ever seen was one that escaped from Antarctica. It was 335 kilometres long and 97 kilometres wide!

Only the smallest part of an iceberg floats above the water. Near Greenland, 167 metres of ice was seen towering above the water. The bottom of this iceberg was about 500 metres below the surface!

Rivers

Flowing waters

Rivers carry water from inland out to sea, or to a lake. Some rivers start off as tiny **streams,** with melting ice and snow from mountains trickling in to them. As they flow across the land, the water cuts its way into the landscape creating **channels** and **gorges**.

Along the length of the river small streams, called **tributaries**, flow into it. They put more water into the river, making it bigger and wider.

Waterfalls

If a river flows down a very steep hill, or over a cliff, a **waterfall** is formed.

Some waterfalls are very high. The highest waterfall in the world is the Angel Falls in Venezuela. Here the water of the Caroni River drops from a height of nearly 1000 metres!

Record rivers

The longest river in the world is the **Nile**. It begins life as two rivers in Africa. The Blue Nile comes out of Lake Tana in Ethiopia. The White Nile comes from Lake Victoria. The Blue Nile and White Nile join up and then flow into the Mediterranean Sea on the coast of Egypt.

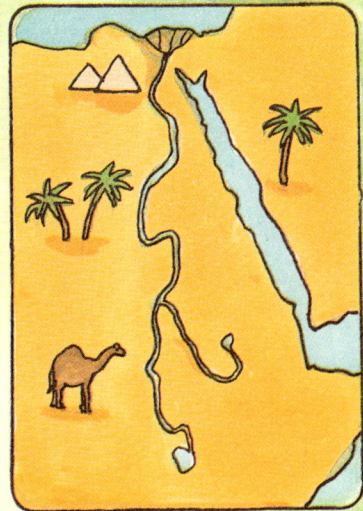

Deserts

Dry places

Deserts are very dry places. They have less than 30 centimetres of rainfall all year. Not all deserts are sandy. Most are rocky areas covered with stones or pebbles.

In sandy deserts, high winds can whip up the sand into **sandstorms**. These can be very fierce and can cause a lot of damage. Winds blow the sand into mounds called dunes, like giant ripples running across the desert.

The Earth's largest desert is the Sahara in Africa. The Sahara is nearly as big as Australia!

Desert life

Plants that grow in the desert must be very tough. They must survive long periods without any rain. American deserts have cacti, all different shapes and sizes. Cacti store water in their stems to use when there is no rainfall.

Lizards and snakes also live in the deserts. Although it can be very hot during the daytime, at night deserts can be very cold. Most of the desert animals only come out at sunset and sunrise. This means that they avoid the heat of the day and the cold of the night.

Forests

What is a forest?
A forest is an area of land covered by trees.

A **coniferous forest** is made up of trees which are **evergreen**. Evergreen means that the trees always have leaves on them.

Temperate forests grow in mild climates. They have trees like oak, elm and beech.

Man-made forest
Man-made forests grow wood for timber and paper. When trees are cut down, new ones are planted in their place. So, natural forests and woodlands can be protected.

Tropical rainforests
Tropical rainforests are found near the Earth's Equator. They are very hot and damp all year through.

These rainforests are very important to life on Earth. The trees and plants use carbon dioxide from the atmosphere and convert it to oxygen. We need oxygen to breathe.

Layers of rain forest
A tropical rainforest has five main levels:

The **attic** is where the tallest trees stick up above the canopy.

The **canopy layer** blocks all sunlight out below. The tops of the trees touch each other so the cover is very thick. Rain sifts through this heavy layer of leaves to become mist as it trickles down to the lower layers.

Younger trees push through the **middle layer** or **understorey**, to get to the sunlight above. Creepers and climbing plants live here, stretching up towards the light.

The **lower layer** is darker. Only shrubs and young trees grow here.

The **forest floor** doesn't see much sunlight. Most plants need light to live but seedlings, fungi and herbs do not. They can survive here on the dark forest floor.

Watching the Earth

Looking Down From Space

The Moon circles planet Earth. We call it a **satellite** -
something which circles something else. The Moon
is our only natural satellite.

A small spacecraft that
moves around the Earth
is called an artificial
satellite. These satellites
don't need pilots.

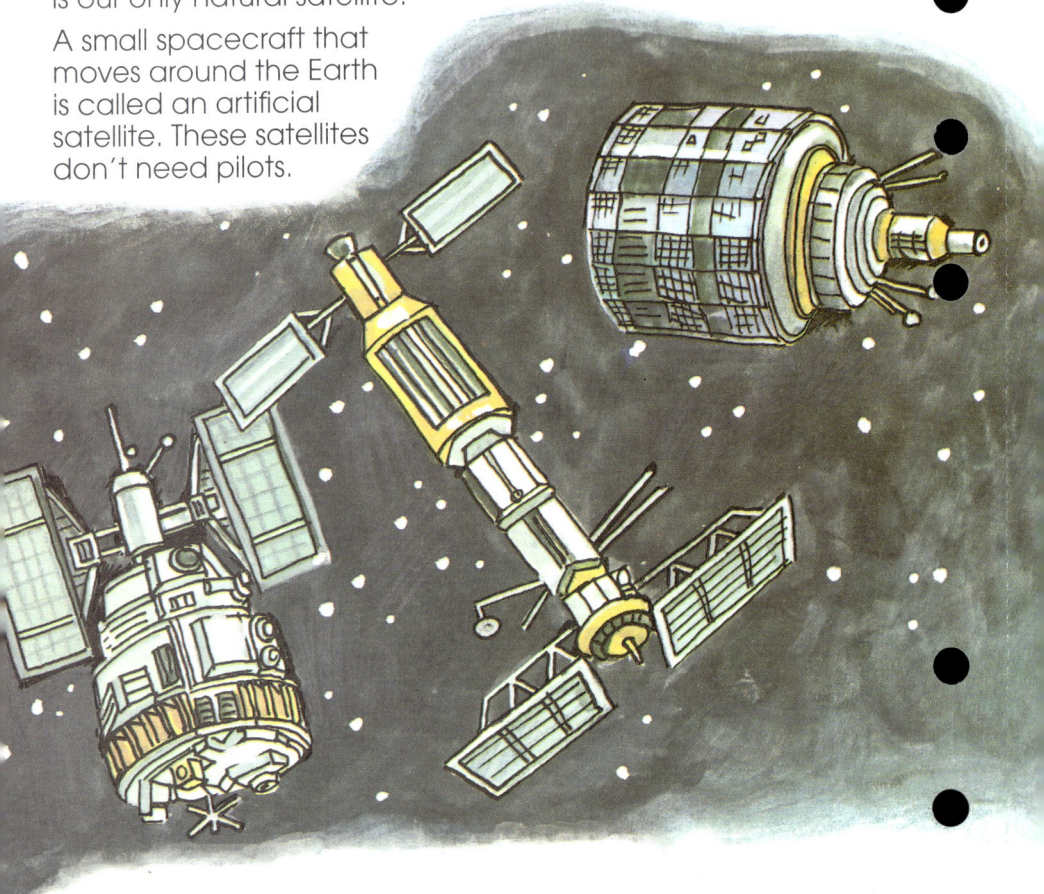

Above the earth, there are many artificial satellites. They
carry special cameras and equipment and do lots of
different jobs. Did you know that, if you telephone a friend
in another country, your voice may be beamed through a
satellite? Satellites are used by telephone companies to
link callers in different parts of the world. Others are used
to transmit television pictures all round the world.

Caring for the world

Scientists use satellites to keep an eye on the Earth. They take pictures of the planet's surface with special cameras on board the satellite. These pictures help us to make maps and to pick out new supplies of oil and gas. They also show us things like forest fires, icebergs and pollution at sea so we can act quickly to warn people who may be in danger.

Satellite photograph

Weather satellites

From high above the Earth's surface, weather satellites can measure the temperature of the air. They can also take pictures showing whole patterns of clouds. The information is sent down to Earth and is used by scientists to forecast the weather. Pictures like this are shown every day on television. Before satellites were used, it was a lot harder to get the weather forecast right!

Day and Night

Day and night

Can you imagine you are up in space, looking down at the Earth? You would see half the planet lit up by the sun. There it is daytime.

The side facing away from the sun is in darkness. Here it is night. The Earth is spinning round slowly, like a top. It takes round 24 hours to complete a spin. As this happens, each place on Earth has day followed by night.

sunrise and sunset

Have you ever been awake early enough to see the dawn? If so, you will have noticed how the sun seems to rise from behind the Earth. Yet the sun doesn't really move up in the sky. Our part of the Earth is actually turning towards the sun, making the sun appear to rise.

At sunset, the sun looks as if it's going down behind the Earth. This is because the Earth is turning away from the sun.

Do-it-Yourself!

You can see how day and night happen for yourself. Stand in a dark room with a ball. This will be the Earth. You might even want to draw an X where your town is.

Now shine a torch onto the ball. This shows the sunlight. You will see that half the ball is lit up, just like the sun lights up half the Earth's surface.

If you spin the globe, you will see that different parts of the Earth have day followed by night.

Words about Our World

atmosphere the air, or gases, that surround a planet

celestial to do with the sky

carbon dioxide the gas we breathe out and plants take in

continents one of Earth's large pieces of land: Asia, Africa, Europe, North and South America and Antarctica

core the inner part of the Earth

crust the outer layer of the Earth

desert an area of very dry land which gets little rain

Earth third planet from the sun; the planet we live on

earthquake a sudden movement within the Earth's crust which causes shock waves which make the Earth's surface shake.

Equator an imaginary line around the middle of the earth

galaxy huge spiral-shaped collections of millions of stars floating in space

islands a piece of land, surrounded by water, that is smaller than a continent

icebergs a large piece of ice floating in the sea

magma the top layer of rock in the mantle

mantle a moving layer of very hot rock under the Earth's crust.

Milky Way the name of our Galaxy

orbit the path of one thing in the sky around another

oxygen the gas we breathe

plates the pieces of the Earth's crust that carry the continents

satellite something, natural or man-made, that travels around a planet in space

solar system solar means 'of the sun'. The nine planets that orbit the sun

star a hot body of gases such as the sun

sun our star, centre of our solar system, a glowing ball of gases

volcano a hole in the Earth's crust where hot ash, pieces of rock, liquid rock and dust spurt out

waterfall where a river falls straight down, over a cliff of steep hill